s o s PLANET EARTH

WATER SQUEEZE

Written by Mary O'Neill
Illustrated and Designed by John Bindon

Library of Congress Cataloging-in-Publication Data

O'Neill, Mary, (date)
Water squeeze / by Mary O'Neill; illustrated by John Bindon.
    p.   cm.—(SOS planet earth)
    Summary: Discusses the importance of water in our lives and the
dangers we create when we pollute the waters of the planet.
    ISBN 0-8167-2080-0 (lib. bdg.)   ISBN 0-8167-2081-9 (pbk.)
    1. Water—Pollution—Juvenile literature.   2. Water—Juvenile
literature.   [1. Water.   2. Water—Pollution.   3. Pollution.]
I. Bindon, John, ill.   II. Title.   III. Series.
TD422.054   1991
363.73 94—dc20   89-77456

Published by Troll Associates, Mahwah, New Jersey in
association with Vanwell Publishing Limited.
Copyright © 1991 by Mokum Publishing Inc.

Printed in the United States of America.

10 9 8 7 6 5 4 3 2 1

Troll Associates

About This Book...

Where would we be without water? All life on earth depends on it. Most of us take water for granted since it covers so much of our planet. But our water is in danger. Many plants and animals that live in water are dying from the poisons dumped into it. Humans also suffer from unclean drinking water. In some parts of the world today, there simply isn't enough clean fresh water for people's basic needs.

In this book you'll learn about the amazing jobs water does for us. You'll also learn about the ways we have abused and polluted water.

There is a lot we can all do to help protect our supply of water on earth. People just like you are working hard right now to save our water for the future. Read on, and find out how you can become part of the great water rescue team!

Contents

Surrounded by Water

Just look around you. You'll see water everywhere. It might be raining today. You might be enjoying a cool drink. You might be sitting beside a lake or an ocean. Or perhaps your town is covered in a layer of snow. You could point to any of these things and say, "There's water!"

Water is also found in many places and things you would never suspect. In fact, water makes up part of every living thing on earth. It's an ingredient in many nonliving things as well. Water is so important that without it, life on earth would be impossible.

WATERMELON
97% Water

MEATS
50-70% Water

TOMATOES
94% Water

BREADS
30% Water

From the Dinner Table to Our Bodies

If you have ever tended a garden, you know that plants need a great deal of water. Since they "drink" so much, it's no surprise that the fruits and vegetables we eat are made up mainly of water. Watermelons and tomatoes are perhaps the wettest foods we eat. They are almost entirely water. Other foods on our dinner table are wetter than you might think. Different kinds of meat are between one-half and three-fourths water. Bread is about one-third water.

It also takes a lot of water to grow these foods for our dinner table. For example, to grow just over two pounds (one kilogram) of dry wheat, about 400 gallons (1,500 liters) of water are needed. If you think wheat is thirsty, imagine how much a cow or pig can drink in its lifetime! It takes an amazing 5,000 to 15,000 gallons (20,000 to 60,000 liters) to produce just over two pounds (one kilogram) of meat from these animals. This includes the water an animal actually drinks, as well as the water needed to grow the food it eats.

Our bodies are mostly water, too. The average person is two-thirds water. Water is found in every cell of our bodies.

Using Water

Besides being a basic part of all plants and animals, water is needed for many of the things we do every day. Our uses for water fall into three main categories: industrial use, irrigation, and domestic use.

Industries that manufacture, or make, things are often located beside a body of water. Water is an important part of many manufactured goods, such as paper and plastics. Water can also be a route for transportation if the finished products have to be shipped to other places. Water might also be used as a source of electricity to provide energy for the machines. Or water can be used to cool or clean equipment and materials used in the manufacturing process.

Irrigation is the process of supplying water to dry land through ditches, sprinklers, or pipes. In areas of the world that don't receive much rain, farmers must depend on irrigation to provide water for their crops.

In dry regions, people must use almost all of their domestic water supply for basic needs such as washing, cooking, and drinking. But in water-rich nations, a great deal of water is used for luxuries such as keeping lawns green.

Flushing It All Away!

People use water differently according to how much they have. In some water-rich nations, over one quarter of all the water used in the home goes down the toilet! But in countries where water is scarce, there may be no flush toilets at all. Human wastes end up as fertilizer in the fields instead of going down the toilet!

Living Without Water

Imagine what life would be like if you awakened one day to find your town's water supply had been cut off. How would your daily life change? How would you wash yourself? What kinds of meals could be cooked? In some areas of the world, people have to face these problems every day.

Today about three-fourths of the fresh water used around the world is for irrigation. But industry is expected to use more and more.

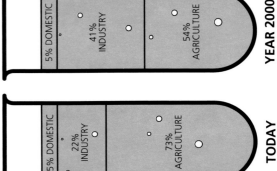

TODAY	YEAR 2000
5% DOMESTIC	5% DOMESTIC
22% INDUSTRY	41% INDUSTRY
73% AGRICULTURE	54% AGRICULTURE

Water—The Miracle Substance

Water is important to life on earth because it behaves like no other substance. A water molecule is formed when two atoms of a gas called hydrogen join one atom of a gas called oxygen. (Molecules are tiny particles made up of even smaller particles called atoms.) Together, the atoms of these two gases make a molecule that has a strong pull, or attraction, for other molecules. This attraction is one reason why water mixes so well with other substances—including pollution.

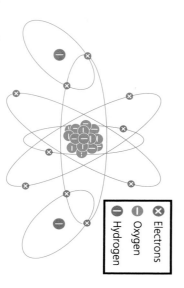

❌ Electrons
❶ Oxygen
➖ Hydrogen

A Great Mixer

If you take a spoonful of sugar and stir it into a glass of water, the sugar seems to disappear. Where does it go? It dissolves in the water. The sugar molecules are so attracted to the water molecules that the sugar joins with the water. Just tasting the water will tell you the sugar is still there!

Water travels far and wide to every corner of the planet. And wherever it goes, water "carries" substances with it.

Fish and other water animals don't breathe as we do, but they need oxygen just the same. Water carries the oxygen fish need in the same way the air carries our oxygen. Plants rely on a gas called carbon dioxide to help them make food. Water provides carbon dioxide to underwater plants and brings minerals to land plants. Water is also home to tiny living things such as bacteria and microbes. They provide food for other creatures and help break down dead matter in the water.

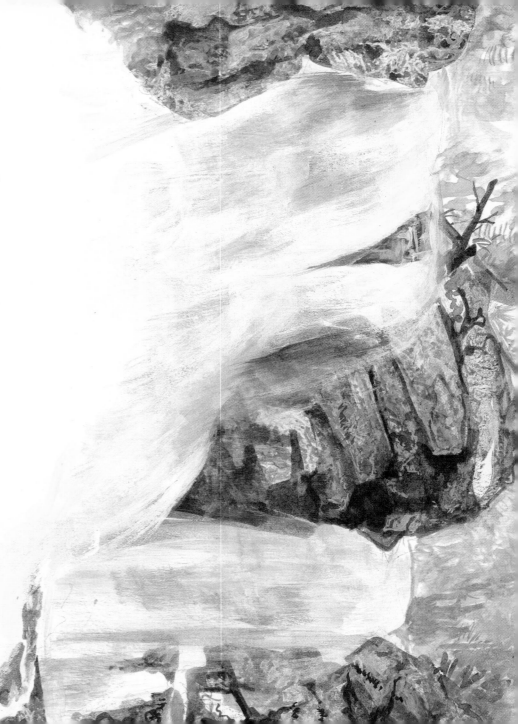

Setting the Temperature

Have you ever cooled yourself on a hot day by splashing cold water on your face? We use water to cool a great many things, from our own bodies to industrial waste. Why is water so good at this cooling job?

Water can be a solid, a liquid, or a gas. The clouds floating overhead are water vapor, or gas. Snow and ice are examples of water as a solid. And of course the water you drink is a liquid. If we want to turn ice into water, or water into vapor, we have to use heat. For example, as ice melts, it takes heat from its surroundings. This is how it cools our drinks. As water turns into vapor, or evaporates, it also absorbs heat. This is how cold water cools our skin. It uses heat energy from our bodies to change into vapor.

Controlling Earth's Temperature

If just a thin film of water can cool our bodies, imagine what effect the oceans can have on earth's temperature. Water helps keep our climate stable.

If you've ever sat by an ocean or lake on a hot day, you've probably felt wonderful cooling breezes. The air over the water cools as the water evaporates. The water molecules take heat from the air in order to turn to vapor. But water's main job isn't to cool things down. Water actually keeps the earth warm!

Water vapor in the air joins together with carbon dioxide, a gas, to form a shield around our planet. This shield traps heat from the sun. Certain rays from the sun pass through the shield but can't escape back into space. We call this the "greenhouse effect." Without this heat trap, the average temperature on earth would be minus 10 degrees Fahrenheit (minus 24 degrees Celsius). Thanks to water in the air, the greenhouse effect keeps the world's average temperature at a comfortable 60 degrees Fahrenheit (16 degrees Celsius). But the greenhouse effect can also cause problems. As more carbon dioxide from pollution enters our air, too much of the sun's heat might be trapped within the atmosphere. Scientists today worry that the earth may grow too warm.

Sweat—The Body's Cooling System

When your body sweats, it's just trying to do what the cold water splashed on your face does—cool you down! The sweat on your skin absorbs your body heat. As sweat evaporates, your skin temperature goes down.

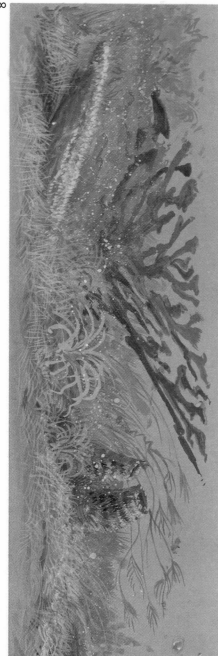

The Water Cycle: What Goes Up...

Where does water go when we pour it down a sink? Are we wasting water? When we speak of "wasting water," we mean usable water. Only a tiny fraction of the water on earth can be used for drinking, washing, or irrigating fields. So when you pour tap water down a sink, you are wasting fresh usable water.

But water does not disappear. It simply changes from one form to another in a continuing cycle. This diagram shows the movement of water between the earth and our atmosphere. We call this process the water cycle.

In the water cycle, water on earth's surface evaporates into the atmosphere as a gas, or vapor. Eventually the vapor in the air comes together to form clouds. As the vapor in the clouds cools, it turns back into a liquid. Then it falls to the earth's surface as rain.

Most rain falls straight into the oceans, which cover seventy percent of the planet. The rain that falls on land either collects in bodies of fresh water or soaks into the soil and is used by plant roots. Eventually the water will rise again as vapor, and the cycle will start over.

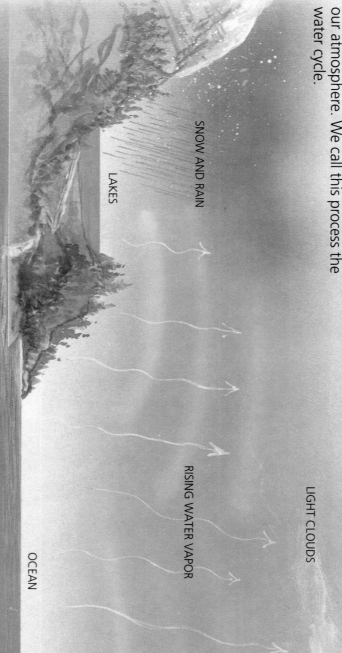

SNOW AND RAIN

LAKES

OCEAN

RISING WATER VAPOR

LIGHT CLOUDS

Earth's Washing Machine

Water picks up a number of "hitchhikers" as it falls from the sky. Some of these substances are harmful gases and chemicals that have been released into the air. Water also picks up minerals and microbes as it washes over the land.

Sooner or later most water on earth ends up in the ocean. You probably already know one main ingredient of ocean water—salt! Salt water isn't good for drinking or watering fields. But many types of plants and animals can make their homes in salt water. Some of these ocean dwellers are too small to see, yet they do a number of important jobs. Many tiny plants and animals live on the gases and chemicals washed into the oceans. By eating these substances, these life forms actually break down some of the gases and chemicals that might poison our atmosphere.

The ocean does its cleaning job in another way. As living plants and animals in the ocean die, many are broken down by bacteria. Whatever can't be consumed settles on the bottom of the sea. As time goes by, this dead matter builds up to form new layers of sea floor.

RAIN CLOUDS

RAIN

Not to Everyone's Taste

Many people like to eat crabs. But nobody would want to share a meal with a crab. These ocean crawlers live off the dead matter that falls to the sea floor. Any matter that's left to rot gives off a poisonous gas called hydrogen sulfide. If too much of this gas filled the oceans, many creatures would die of poisoning. But nature doesn't waste

anything. Hydrogen sulfide provides food for a certain kind of bacteria! And as this bacteria breaks down hydrogen sulfide in the water, it also provides food for other ocean creatures. In 1977 scientists discovered a whole new community of worms and crabs living deep in the ocean. These bottom feeders depend on the hydrogen-sulfide eaters both for food and for poison-free water!

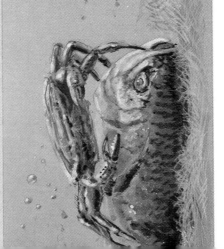

Our Freshwater Supply

With so much water covering the earth, why worry about saving water? Most of earth's water is salty ocean water that can't be used for drinking or irrigation. Only three percent of all water is fresh, and most of this is frozen at the North and South Poles. Water we can actually use makes up less than one percent of all the water on earth. But even this one percent should be more than enough for all our needs. It represents one million cubic miles (4.3 million cubic kilometers) of water!

Our two main worries about water are pollution and distribution. Some areas of the world have plenty of fresh water, but the sources of this fresh water are becoming unsafe. In other regions people have trouble simply getting the water to where it's needed. This map shows the regions that have the greatest need for water.

THE WORLD'S WATER

USABLE FRESHWATER 1%

FROZEN FRESHWATER 2%

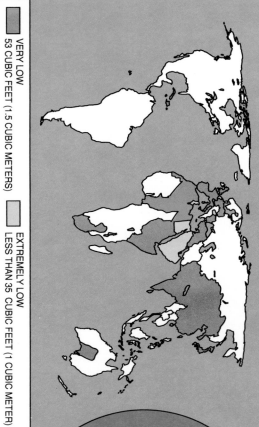

WATER AVAILABILITY PER PERSON - PER YEAR

VERY LOW
53 CUBIC FEET (1.5 CUBIC METERS)

EXTREMELY LOW
LESS THAN 35 CUBIC FEET (1 CUBIC METER)

When Is Water Polluted?

Water is almost never pure. Even clean rainwater carries oxygen, nitrogen, and carbon dioxide. Water might also carry dust, salts, acids, and other impurities it has picked up in the air. So what do we mean when we call water polluted? Water is said to be polluted when it carries so many other substances that it harms life.

Pollution comes in many different forms. It may be germs that cause disease, poisonous chemicals, or too many minerals and nutrients. Water pollution usually occurs when something outside the usual water cycle steps in to disturb the balance of life.

Decade of Water

Because safe water is so important, the United Nations named 1981-1990 the International Water Supply and Sanitation Decade. During those years thousands of projects around the world aimed to bring clean, safe water to where it was needed. The International Water Supply and Sanitation Decade did not meet all its goals. But for the first time millions of people received water pumps and water education.

A New World of Industry

Until the nineteenth century most people spent their lives on farms. Communities were small and spread out. The waste from these communities did little harm to the environment.

In those days most products were made by hand. Clothes were sewn by hand, with some leather goods sewn by a foot-pedal sewing machine. Blacksmiths forged metal tools. Carpenters used simple hand tools to create wood furniture. These methods did not produce many dangerous wastes.

Everything changed as marvelous new machines were invented. The steam engine came along to replace sources of power such as windmills. Now machines could run continuously as long as they had fuel. Machines powered by steam engines could produce things faster, cheaper, and in greater numbers than ever before. New spinning and weaving machines brought great improvements to clothing production. Steamboats and railway engines made transportation faster. And new methods of metalworking meant that farmers had stronger steel tools to replace brittle iron ones.

From the Country to the City

One of the effects of the new industries was to bring millions of people together. As factories sprang up, many people flocked from the countryside to form towns nearby. They were eager to earn money in new jobs. Eventually towns swelled to become overflowing cities. Many of these cities grew so fast that there wasn't time to plan for proper housing or sewage systems.

Where did all the garbage go? Most of it went straight into rivers and streams. Many people thought of these waters as a bottomless pit that could swallow any amount of waste. The early sewage systems simply carried waste from houses and gutters and dumped it into water nearby. Often the waste water dumped upstream was used downstream for washing and drinking. Terrible diseases broke out.

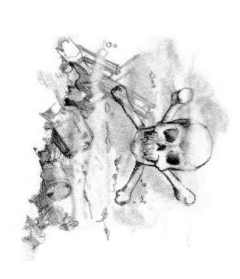

Factories on the Water

Household sewage wasn't the only filth going into the water. Factories spilled poisonous metals and chemical wastes into the rivers and oceans. Since the water seemed so vast, no one gave much thought to what effects this pollution might have.

Some Harmful Wastes of Industry

Most businesses obey local, state, and federal laws limiting what they can put into the water. But some businesses don't. Lakes, streams, and ground water (water stored beneath the earth's surface in the space between soil and rock formations) can all be fouled and even poisoned by an industry's illegal dumping. Sometimes pollution can also be an accident, as in the case of a broken container that leaks waste.

Accident or not, pollution always hurts. The chart below lists the main types of pollution that can come from industries. Each of these metals and chemicals poses serious risks to human health. Phenols and cadmium chloride are suspected of causing cancer. Mercury and lead can damage the nervous system. And chlorinated organic compounds can build up in the body to damage the liver and attack the immune system (the body's defense against disease). In all, more than 700 chemicals have been found in drinking water.

	CHLORINATED ORGANIC COMPOUNDS	MINERALS & OILS	PHENOLS	NITROGEN	PHOSPHORUS	MERCURY	LEAD	CADMIUM
PESTICIDE INDUSTRY	☠	☠			☠	☠		
PETROLEUM REFINING		☠	☠	☠				
TEXTILE INDUSTRY	☠	☠		☠				
PETROCHEMICAL INDUSTRY		☠	☠	☠	☠	☠	☠	
IRON/STEEL INDUSTRY	☠	☠	☠	☠		☠	☠	☠
PULP/PAPER INDUSTRY		☠	☠	☠				
METAL/METAL PLATING INDUSTRY	☠	☠	☠	☠	☠	☠	☠	☠
FERTILIZER INDUSTRY	☠	☠		☠	☠	☠	☠	☠
CHEMICAL INDUSTRY	☠	☠	☠	☠	☠	☠	☠	☠

The Water We Drink

When we want a drink of water, most of us just turn on the faucet. We rarely think about where the water comes from, how it has been treated, or how it gets to our sinks. In days gone by, drinking water was taken straight from rivers, streams, and underground wells. But in most parts of the world today, water has to be taken from carefully chosen sites. Water may also go through some form of treatment to make sure it doesn't contain anything harmful. Even so, many people are concerned with the safety of our drinking water.

Freshwater Sources

Most drinking water comes from lakes, rivers, ground water, and reservoirs. Rainwater that soaks into the earth and then becomes trapped above a solid layer of rock can form aquifers. Aquifers are great underground reservoirs that may be the main source of fresh water in some areas. In the United States, for example, almost half of the population gets its drinking water from aquifers.

All of these water sources can be easily polluted by sewage or deadly wastes. These wastes may seep into the water if they are stored in the ground. Or they may be dumped directly into the water. Because of pollution risks, much of our drinking water is treated with chemicals before it finds its way to our faucets.

HOW DOES WATER GET TO YOU IN A LARGE CITY?

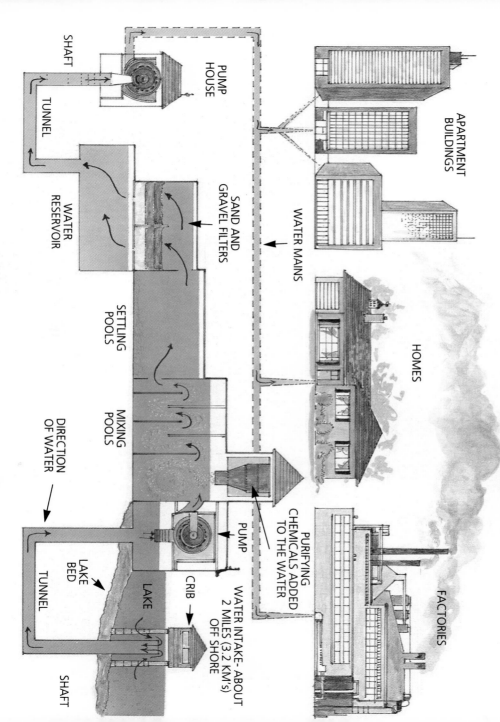

SHAFT

TUNNEL

PUMP HOUSE

WATER RESERVOIR

SAND AND GRAVEL FILTERS

APARTMENT BUILDINGS

WATER MAINS

HOMES

SETTLING POOLS

MIXING POOLS

DIRECTION OF WATER

PURIFYING CHEMICALS ADDED TO THE WATER

PUMP

FACTORIES

LAKE BED

TUNNEL

LAKE

CRIB

WATER INTAKE- ABOUT 2 MILES (3.2 KM's) OFF SHORE

SHAFT

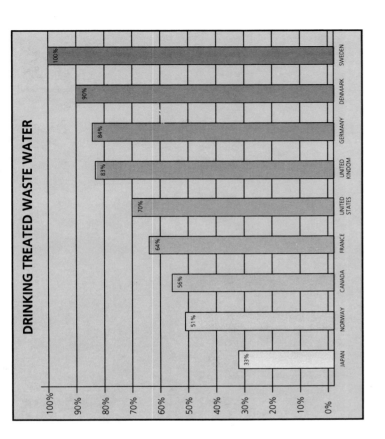

DRINKING TREATED WASTE WATER

Country	Percentage
SWEDEN	100%
DENMARK	90%
GERMANY	84%
UNITED KINGDOM	83%
UNITED STATES	70%
FRANCE	64%
CANADA	56%
NORWAY	51%
JAPAN	33%

Treating Our Water

In crowded or industrial areas, many people have no choice but to drink water that was once heavily polluted with sewage and chemical wastes. In these regions water is treated to make it safe to drink.

Modern water treatment can involve up to three stages. First, solids are removed by allowing them to settle to the bottom of water tanks. Second, bacteria are used to break down any dead matter. In the third stage chemicals such as chlorine and ozone may be used to kill any dangerous bacteria in the water.

Many people are worried that this third treatment stage may do more harm than good. Some studies have linked increased disease rates with the drinking of chemically treated waste water. But with few clean water sources available, many countries have no choice but to recycle waste water by treating it with chemicals.

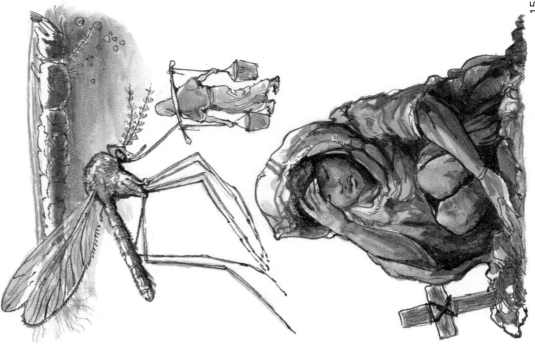

Dirty Water

The world's deadliest killer may be dirty water. In many areas of the world, untreated drinking water is taken from the same water that serves as sewage dumps. In sections of Africa, Southeast Asia, and South America, only two out of five people have safe drinking water. Many other people risk infection with every drop they drink.

Many diseases come from insects that live and breed in the water. Malaria, a severe disease that infects 160 million people annually, is carried by mosquitoes that breed in water. Simple diarrhea, which kills six million children each year, is often caused by drinking water that has been polluted with human waste. Millions more die each year from diseases caused by bacteria in dirty drinking water. Sadly, many of these diseases could be prevented with two simple measures: (1) bring in drinking water through protected pipes, and (2) remove waste water in separate pipes.

Chemical Soup: The Rhine River

WESTERN EUROPE

SWITZERLAND

The Rhine River flows from Switzerland, through Germany and the Netherlands, and empties into the North Sea. The Rhine flows past one fifth of all the world's chemical factories. But it also provides drinking water for twenty million people in Europe! Some countries along the Rhine have no choice but to get their drinking water from this source.

Eyewitness Report

Basel, Switzerland:
November 1, 1986

Early this morning fire raged through an industrial plant on the Rhine River in Basel, Switzerland. Authorities believe that up to thirty tons of toxic chemicals may have been flushed into the river as firefighters battled the blaze. The fire is thought to have been caused by animals chewing through electrical wires. Local officials fear the chemicals will spread down the Rhine River and cause extensive damage to drinking water and water life.

A Lifeless River

In only two hours on November 1, 1986, the Rhine received more pollution than it usually does in one year. Pesticides, dyes, and weed killers from this disaster wiped out river life up to 120 miles (190 kilometers) downstream. When officials tested the river, they unexpectedly found dozens of other substances that had not come from the November 1986 disaster. During the next month other companies confessed to twelve other spills they had never reported.

The effects of this chemical soup were disastrous. Drinking-water supplies over a wide area were threatened. Close to a half-million fish died immediately. Smaller forms of life were also killed, cutting off the food supply for any surviving fish. Scientists said it would take at least ten years for life in the Rhine River to recover.

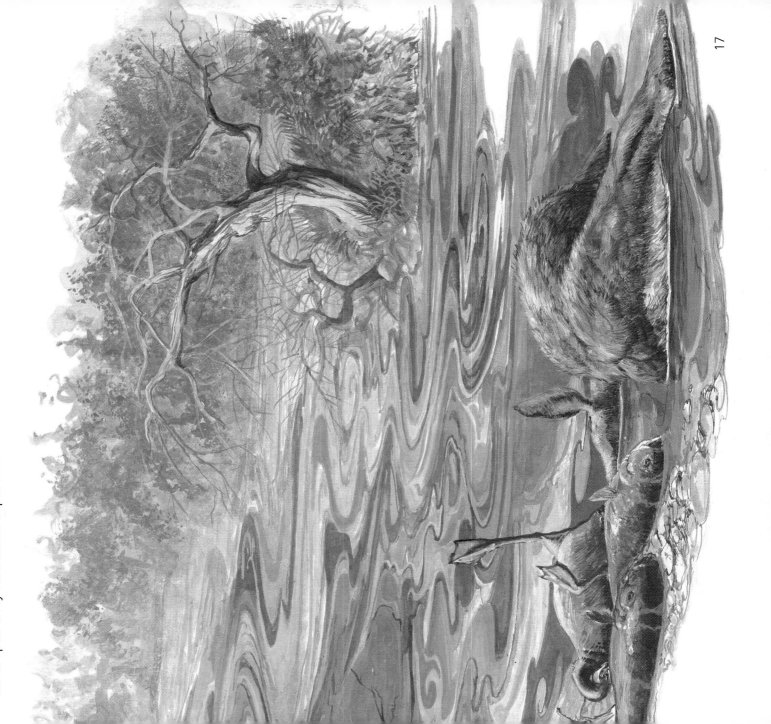

Changing the Balance of Life in the Water

We see only a tiny number of the plants and animals that live in lakes and streams. We're of course familiar with fish, frogs, otters, and water lilies. But perhaps even more important to the life of the water are billions of tiny organisms. They are the bottom link in the chain of water life.

The smallest plants in the water are called phytoplankton. Phytoplankton may be as small as a single cell. But they are able to live on sunlight, minerals in the water, and carbon dioxide. They take in these simple substances and are themselves food for other kinds of life in the water.

Zooplankton are the tiniest animals in the water. Many feed on the phytoplankton. These zooplankton, in turn, become food for larger animals. This chain of "eating and being eaten" continues all the way up to larger animals such as the fish-eating otters. Each member of this "food chain" is important to all the other members. If one member dies off or grows out of control, all the others may be in danger. The balance of life can be easily disturbed by human activities.

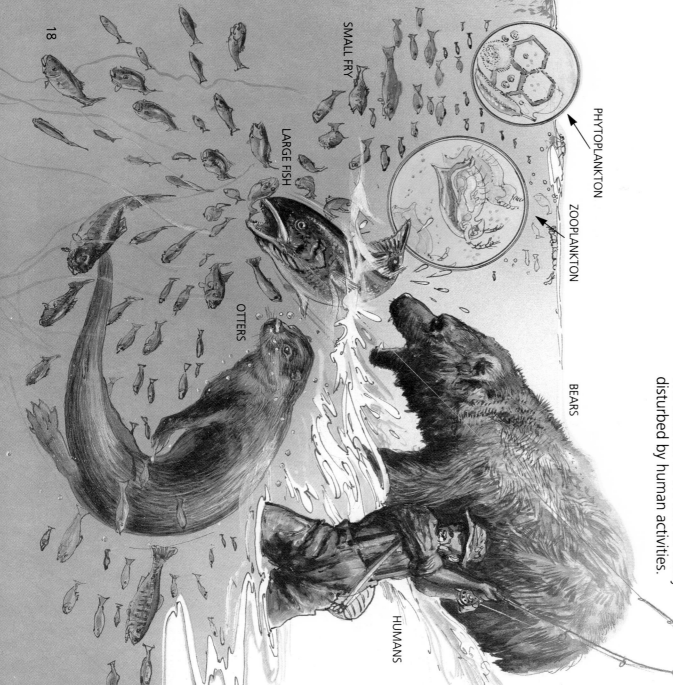

SMALL FRY

PHYTOPLANKTON

ZOOPLANKTON

LARGE FISH

OTTERS

BEARS

HUMANS

Detergent Dump

In the 1950s fishermen began to notice that fish were getting harder to find. Many lakes and streams were covered with a thick green film—a type of phytoplankton called algae. The water underneath was nearly lifeless. What had killed off life there?

It was clear that the algae were doing too well. They were choking off all other life forms. Scientists found that the algae were getting too much food. But where was it coming from?

The substance that seemed to be too plentiful was a group of nutrients called phosphates. Used in both detergents and fertilizers, phosphates were being dumped in sewers by washing machines in the home and by industries. Run-off water from farms was also carrying fertilizer with it.

With so much food, the algae grew out of control. When the algae finally died and sank to the bottom, bacteria that break down rotting algae used up the water's oxygen. And as more oxygen was used up, fish and other animals couldn't breathe.

Captain Conservation: A "Water-Friendly" Laundry Powder

Today many detergents no longer contain phosphates. Their boxes are labeled "phosphate-free." If you want to be sure that doing your laundry isn't harmful to water life, you can make your own "water-friendly" laundry powder!

Add one-third cup of washing soda (sodium carbonate) to your laundry water. When you've put your clothes in the water, add one-half cup of pure soap. That's all it takes to clean a normal-sized load of dirty clothes!

Chain of Poison

Eyewitness Report
Minamata, Japan:
1968

The 3,500 villagers who live in this tiny fishing community are coming to terms with a dreadful tragedy. For years people were silent about the disease that has ravaged their village. Now they are finally speaking about the pollution that has caused their agony.

For almost two decades the villagers of Minamata have suffered from mysterious illnesses. Children were born deformed and brain damaged. Many died. Adults lost control of their body movements or went blind. Many died without knowing the cause of their suffering.

Recently, however, a large chemical company on the shores of Minamata Bay admitted it had been dumping chemical wastes into the bay for years. One of the metals discharged, mercury, found its way into the fish and shellfish eaten daily by the local people. Mercury is known to attack the central nervous system and can deform unborn children.

The chemical company said it stopped discharging mercury into the bay in 1966. But the damage is done. The mercury dumped over nearly thirty years has found its way into the food chain in Minamata Bay.

MINAMATA

Mercury in Water Around the World

The poisoning of Minamata Bay became so infamous that many people now call the disease caused by mercury poisoning "Minamata disease." By 1983 more than 300 people in the bay area had died from it. Nearly 1,500 local people had the disease.

This little bay on the Japanese coast is not the only spot in the world where there are concerns about mercury poisoning. Native Indians in the Canadian provinces of Quebec and Ontario are also worried about mercury levels in their local fish. Dangerous levels of mercury have even been recorded in such unlikely places as Greenland, which is far away from most industries.

At one time small amounts of mercury did not worry people. But the tragedy of Minamata showed that even low levels of this metal—and other chemicals—can slowly build to dangerous levels in water plants and animals.

Toxic Buildup

Harmful chemicals can build up to toxic levels as one form of life feeds upon another. Chemicals are first taken in by the lowest levels of the food chain, the phytoplankton and zooplankton. As these are eaten by small fish and shellfish, the chemicals are stored in the larger animals' body tissues. Their bodies now may have higher levels of chemicals than the water around them! When the fish and shellfish are eaten by larger animals, such as trout, otters, and seals, the chemicals are absorbed into their body tissues at even higher levels. The higher up the food chain, the greater the level of chemicals a creature is likely to carry. It is this deadly chain that may have killed thousands of seals off the coasts of Europe in the summer of 1988.

TOXIC BUILDUP

Eyewitness Report
Anholt, Denmark: May 1988

Local residents report finding hundreds of dead seals on the shores of this tiny island off the coast of Denmark. Both adult seals and pups appear to have died offshore. Scientists are searching for the cause of this mysterious plague.

Den Helder, Netherlands: July 1988

Death appears to be spreading to the offshore mainland seal colonies from rotting seal bodies. The hundreds of bodies has horror to the shoreline of miles has spread of this disaster along But scientists has Yet cause for pollution may suspect been found. have suspect played that water a role.

No definite North Sea cause for pollution been found.

Rosslare, Ireland: September 1988

Bodies of both harbor and gray seals have now been sighted on the eastern coast of Ireland. Since it was first detected in North Sea water, the seal plague has spread as far as the North Atlantic Ocean and the Baltic Sea. Last month scientists finally discovered that a virus is the cause of the seal deaths. PCB pollution is believed to have weakened the seals' immune systems, leaving them easy prey for the deadly virus.

WESTERN EUROPE

NORTH SEA

SEAL DEATHS

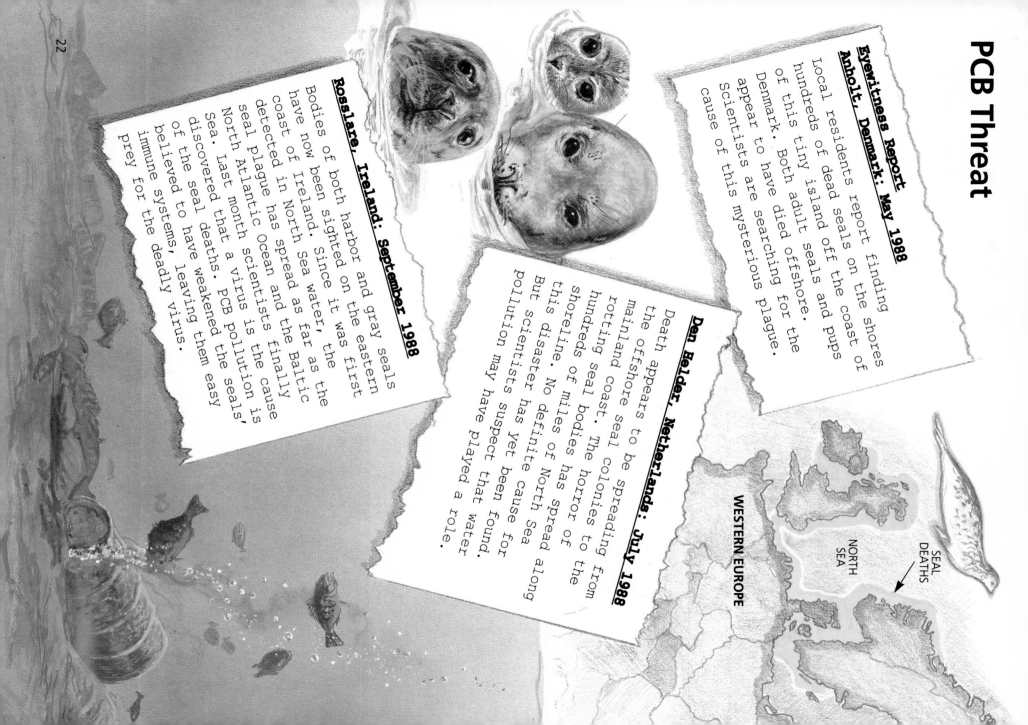

Plague in the North Sea

The North Sea has long been a dumping ground for many European industries. In the early 1960s the North Sea was found to be the first international body of water to be polluted with PCBs, or polychlorinated biphenyls (pol-ee-KLOR-uh-nay-tid by-FEEN-uls). These are industrial substances that can build up to dangerous levels in animals. Animals living in these waters have among the highest levels of PCBs in the world.

In the summer of 1988 mother seals gave birth to their pups too soon. The pups died because they were too immature to survive outside their mothers' bodies. Adult seals suffered from failures of their lungs and other organs. These are all problems scientists have connected with PCBs. Altogether, 12,000 out of a total of 18,000 seals are thought to have died in the region.

PCBs were used extensively in the 1960s to make electrical equipment. Today most countries have stopped making PCBs. But the PCBs produced years ago have not gone away. They do not break down easily. Almost one third of all the PCBs in the world are believed to have leaked into the environment. The rest of the PCBs are either still in use or in storage.

Studies on certain types of deep-ocean mammals show that their bodies have high levels of PCBs—even higher than animals that live on land near PCB sources. Some scientists fear that many more ocean animals may die in the future if PCB pollution continues.

PCB Route to the Oceans

Almost all PCBs in the ocean came down in rainfall. This was a frightening discovery. It means that these substances can find their way into the atmosphere and travel many miles before falling again. Even faraway areas such as the northern Arctic have been hit by PCBs. Polar bears there are in serious trouble because seals make up a large part of their diet. If the seals' bodies contain high levels of PCBs, the polar bears will store even higher levels in their own bodies.

Toxic-Waste Animals

When materials contain so much of a toxic substance that they are dangerous to handle, they are labeled "toxic wastes." Blue-white dolphins off the coast of Europe have been tested for PCB levels. Their bodies were found to have more than sixteen times the amount of PCBs needed to be labeled as toxic waste!

Oceans at Risk

Just how much waste can the oceans take? Each year thousands of tons of harmful substances such as radioactive waste, fertilizers, and pesticides make their way into the ocean. We send nuclear-powered ships and huge oil tankers across the water. When accidents occur, the oceans must swallow their dangerous fuel and cargo. There is a limit to how much we can dump in these waters. And the poisons seem to find ways to haunt us back on land.

Oil's Effects on Ocean Life

Animals caught in oil spills face death from many different causes. Creatures such as otters and birds can freeze to death when oil sticks to their fur and feathers. Smaller creatures such as shellfish, plankton, and microbes may be the first to take in the poisons. But when the oil is then passed through the food chain, the poison may reach creatures far away from the scene of the accident. For example, bears eating salmon out of distant inland rivers may take in poisons the fish have brought from the ocean.

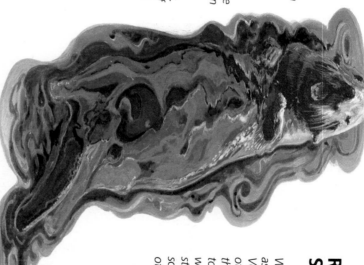

Rolling Up Their Sleeves

On the Alaskan coastline near Prince William Sound, people pitched in to save as many animal victims as possible. Volunteers wrapped freezing birds and otters in towels and blankets to keep them warm, then carried the animals off to be cleaned—often with simple dish-washing liquid! The rescue efforts were still not easy. It can be quite a struggle to scrub a ninety-pound (forty-kilogram) otter that has sharp teeth and claws!

Oil Spills Around the World

In spite of the great damage it caused, the March 1989 tanker spill in Alaska was not the worst oil disaster at sea. Here are just some of the largest tanker spills that have taken place:

1. July 19, 1979: A collision between the Atlantic Empress and the Aegean Captain off the coast of Trinidad and Tobago. **Ninety-two million gallons (350 million liters)** spilled.

2. August 6, 1983: The Castillo de Bellver burns off the coast of South Africa at Cape Town. **Seventy-seven million gallons (292 million liters)** spilled.

3. March 16, 1978: The Cadiz runs aground off the northwest coast of France. **Sixty-eight million gallons (258 million liters)** spilled.

4. March 18, 1967: The Torey Canyon runs aground at Lands End, England. **Thirty-seven million gallons (140 million liters)** spilled.

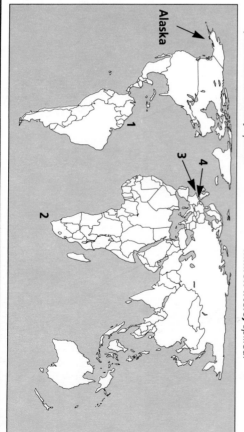

Alaska

Dumping Grounds

Oil-tanker accidents make for only a small part of ocean pollution each year. Our oceans actually face more danger from the millions of tons of waste we dump into them on purpose.

As nations run out of space on land, they are looking toward the oceans as dumps. Even though ocean dumping of heavy metals and cancer-causing wastes is banned, it is legal to dump wastes that contain tiny amounts of these substances. Unfortunately, small traces of toxic wastes can soon add up to tons.

Some countries also permit toxic wastes to be burned at sea in special furnaces called incinerators. Many difficult wastes can be broken down only at high temperatures. PCBs, for example, won't break down until they reach the high temperature of 2462 degrees Fahrenheit (1350 degrees Celsius). Some people feel that incinerators may be the answer to our waste problems. If properly done, the burning removes 99.99 percent of the waste. But for every 10,000 tons burned, one ton is still sent into the air as poisonous ash. Most of the ash falls back to the ocean around the incinerator, where it finds its way into the food chain.

Eyewitness Report

Prince William Sound, Alaska:
March 30, 1989

One of the worst oil spills in U.S. history took a heavy toll in just a few hours on March 24, 1989, when an oil tanker ran aground. This one of the horrible events began on a reef near off the Alaskan coast, a chain of known to scientists as one breeding Prince William Sound. This area is most important rare forms of ocean grounds for many feeding forms of life.

In its hull, the tanker took a deadly cargo (forty-four holes in its gallons oil life. With poured out a few deadly cargo million of crude oil 800 tanker eleven liters) More than all, million into the water) coast one million kilometers) of dying spewed (1,287 in oil. Dead and bodies miles covered sea birds, washed up on were and sea otters, tried to save otters coated in oil, as possible. were shores. Rescuers as possible local shores victims concerned the as many of the are one of The local fishermen is one of salmon Local this water salmon young because largest just as young salmon world's occurred just as from spill occured to be released. But the were due hatcheries scientists as commercial hatcheries of the oil and young greatest concern effects of the young the long-term by plankton and it is taken in by plankton fish.

Floods and Droughts

In much of the world, the amount of rainwater is the biggest problem. When too little rain falls, we say that an area is in a drought. Too much rain at once may cause terrible floods.

In the past we thought of droughts and floods as disasters that were brought on by nature alone. Now we know that poor treatment of soil and plant life can cause land to dry up. Clearing land near water routes can cause flooding, because trees and plants around rivers normally act like a wall to hold back water when heavy rains fall. When people allow the rich soil of entire regions to wash away, or erode, deserts can slowly develop.

Trees: The Water Anchors

Trees do a lot more than provide shade and beauty. They play an important role in balancing the amount of water held by the soil and atmosphere. Below ground, trees have great root systems to help prop them up. These roots suck up water and nutrients from the soil, so rain does not simply run off or evaporate.

Trees also give back some of the water they drink. They release vapor from tiny openings in their leaves. This means that over a wide forested area, there is a steady cycle of rain coming down and water vapor going back up. Rainfall can make its way slowly into the ground water, lakes, and streams. This reduces the chances of flooding or soil erosion.

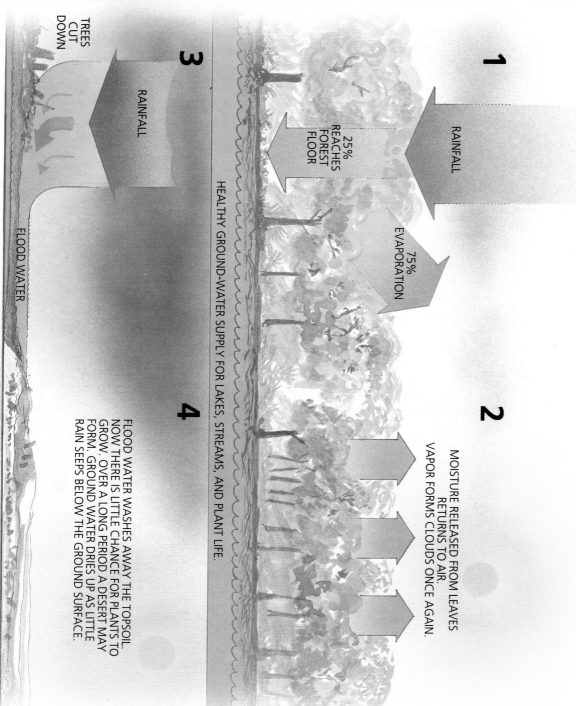

1

RAINFALL

25% REACHES FOREST FLOOR

75% EVAPORATION

2

MOISTURE RELEASED FROM LEAVES RETURNS TO AIR. VAPOR FORMS CLOUDS ONCE AGAIN.

3

RAINFALL

TREES CUT DOWN

FLOOD WATER

HEALTHY GROUND-WATER SUPPLY FOR LAKES, STREAMS, AND PLANT LIFE.

4

FLOOD WATER WASHES AWAY THE TOPSOIL. NOW THERE IS LITTLE CHANCE FOR PLANTS TO GROW. OVER A LONG PERIOD A DESERT MAY FORM. GROUND WATER DRIES UP AS LITTLE RAIN SEEPS BELOW THE GROUND SURFACE.

Nothing to Hold On to

In many parts of the world today, forests are being cut down to supply wood or to make way for ranches and farms. What effect does this have on the land?

First, there may be flooding. Without trees to act as sponges, rainwater simply runs into nearby lakes and rivers. If too much rain falls at once, lakes and rivers may overflow. Terrible floods can wipe out nearby farms and villages.

Without trees, rain also "steals" the precious layer of ground called topsoil, stripping the land of the nutrients plants need to grow. Without plant life, the land grows drier and drier. After many years rich soil may turn to desert.

Saviors of the Forest

In 1974 some women in northern India heard their local forests were to be chopped down. To protect their trees, the women sat in the forests night and day. They threatened to hug the trees if loggers came near them. The protest worked! The women saved almost 5,000 square miles (12,000 square kilometers) of forest. Their efforts spread throughout the region, and people have successfully protected many more forests in India.

Captain Conservation: Be "Water-Smart" About Lawns and Gardens

Lawns and gardens need about one-fifth inch (five millimeters) of water per day during warm weather and less than that during cool weather. Water lawns and gardens every three to five days rather than for a short period every day. Apply one-fifth inch of water for each day since last watering in warm weather. Do it during the cool part of the day, either morning or evening.

And keep grass at a relatively tall height of two or three inches (five to eight centimeters).

Killing Rains

In the forests of Canada the sugar-maple harvest slows to a trickle. The faces of ancient statues in Greece are slowly being eaten away. Four thousand lakes in Sweden no longer have any fish. More than half of all the woodlands in Germany are dead or dying. Around the world a deadly rain of acid is believed to be slowly eating away at life in many forests and lakes.

Acid rain begins with the gases sulfur dioxide and nitrogen oxides. Sulfur and nitrogen enter the atmosphere when industries burn coal and oil. Most nitrogen oxides come from car exhaust. Winds may blow these acids hundreds of miles from their source. When it rains or snows, the acids fall into lakes and forests below.

Death Underwater

Acid lakes can be almost lifeless. A filmy layer of algae might cover these waters. But under the surface, the chemistry of the lake has changed, making life almost impossible for plants and animals.

Scientists believe it may not be the acid itself that's the killer. Instead, the acid in the water increases the levels of metals such as mercury, aluminum, lead, and others that are normally present. Of these metals, aluminum is the deadliest for fish. Aluminum builds up in their gills, lowering the amount of oxygen getting into their bloodstream. Over time the fish die from lack of oxygen.

Yet the first to be affected by acid rain are bacteria. They are needed to break down dead matter on the bottom of lakes. The dead matter can then release rich minerals and nutrients that life in the lake depends on. Without the bacteria, dead plants just pile up on the lake bottom.

The eggs of fish and other lake animals are also destroyed by acid. With no new creatures hatching, the cycle of life in the water comes to an end.

28

Bare Forests

The effects of acid rain on forests have long been a mystery. As thousands of acres of woodlands disappeared, scientists at first couldn't explain how the damage was done. Now they believe acid rain affects forests in three different ways: by removing nutrients from leaves, by taking nutrients from the soil, and by changing harmless aluminum in the soil into toxic forms.

Captain Conservation: Turning Off the Acid

Car exhaust is one of the biggest causes of acid rain. You can help decrease the acid by taking fewer car trips. The next time you need to go somewhere, take public transportation if you can. Or better yet, why not walk or ride a bicycle?

Cleaner Smokestacks

New laws are forcing industries to put "scrubbers" in their smokestacks. Scrubbers are devices that collect chemicals on their surfaces, so acids won't escape into the air. Scrubbers won't bring back dead trees or lakes, but they can save others from being destroyed.

Cleaning Up at Home

You might think water pollution is caused only by industries or by sewage pipes emptying into rivers and oceans. But the problem often starts right in our own homes. A number of products we use are harmful to the environment. We often flush chemicals down the drain without thinking where they are going.

In the garden we use pesticides to keep bugs off plants and herbicides to kill weeds. These chemicals soak into the ground and can end up in streams and rivers. The fertilizers we use to make the lawn green also find their way into nearby water. There they can cause water plants to grow out of control.

Today many products are available that don't pollute the water or damage the environment. As you already learned, some detergents are phosphate-free. Here are more ways you can clean your home, fight insect pests, and control weeds in ways that are "water-friendly." Sometimes these "water-friendly" ways take more work. But they will all make sure we have safe, drinkable water for a long time to come.

Air Fresheners

To keep your house smelling nice, use small bags of dried flowers and herbs instead of ready-made air fresheners. House plants are even better, because they do more than just cover bad odors. They actually clean the air.

For Sparkling Dishes...

Try using pure soap. A little vinegar in the water will help take the grease off dishes.

Painless Weeding

A garden doesn't have to be completely free of weeds to be healthy. Some wild plants are actually quite beautiful. And many types of weeds are good for the soil. But if weeds seem to be choking your garden, you can stop them from growing with "plant collars."

As you plant each seedling, place a circle or square of thick paper around its base. Leave a hole in the paper just big enough for the plant's stem to grow through. Your pieces of paper should be large enough to cover most of the soil between plants. The paper will block sunlight from any weeds trying to grow underneath. Weeds that grow between seedlings can be removed by hand.

Keeping Pests Off Your Plants

Instead of a chemical pest killer, try a soap mixture on your garden. Insects don't like the taste, and soap doesn't harm plants. For each quart (liter) of water, mix in two tablespoons (thirty milliliters) of pure liquid soap. Spray on growing plants. Of course, wash any vegetables before you eat them.

Be a Friend to Water

There is no magic solution that will clean up all our water at once. But when many people act together, each doing a little bit, great things can happen! Here are some simple things you can start doing today to help protect and save water:

* **Don't waste water. Remember, water might be all around us, but only a tiny bit is fresh.**

* **Don't waste other things. Think about all the steps it takes to make the things you own. Many of these steps create pollution.**

* **Try not to use products that have dangerous chemicals in them. Many of these chemicals wind up in our water sooner or later.**

* **If you live in a dry area, help out with a tree-planting project.**

* **Get to know what plants and animals live in the streams, lakes, ponds, rivers, or oceans near you. Find out what each needs for a healthy life and how it helps other life in the water. There may come a day when these neighbors will need help from you!**

Index